W9-BMK-378

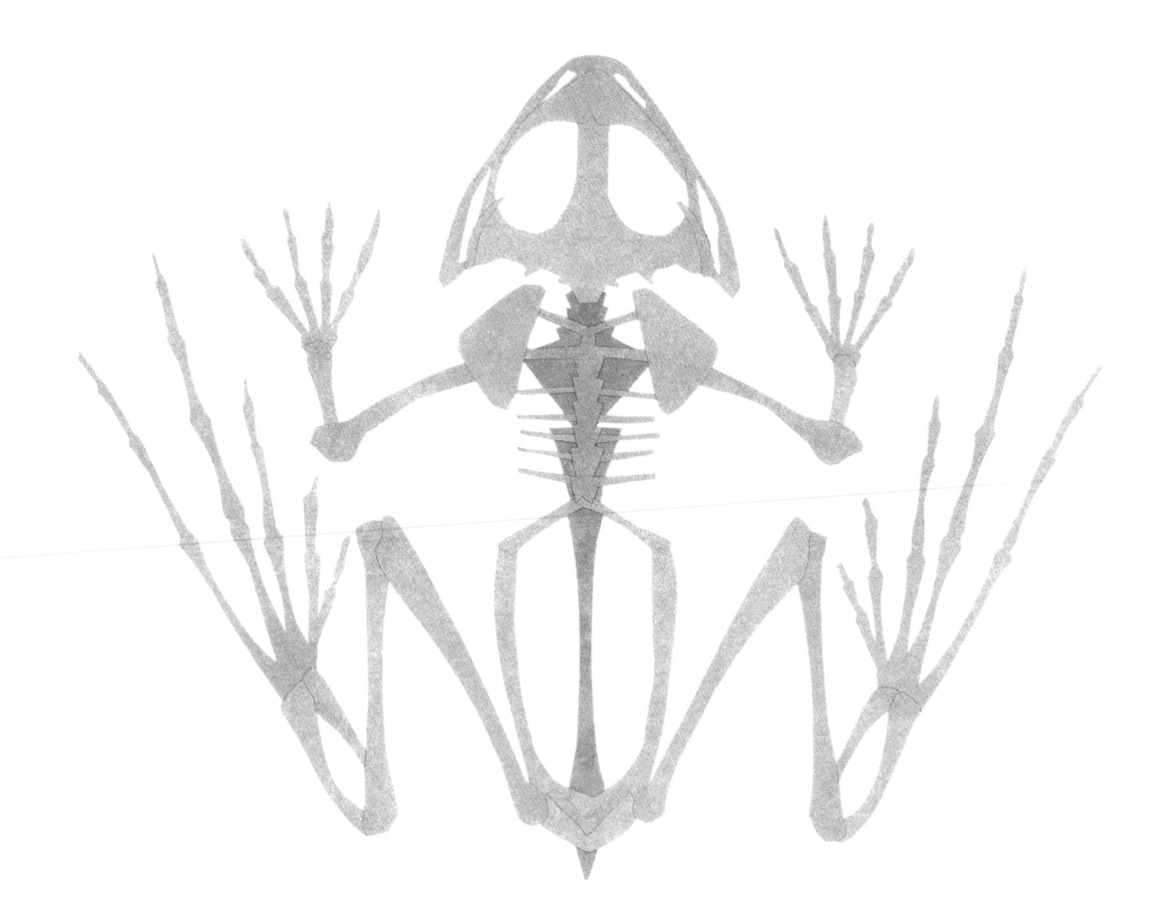

For Jamie

hmhco.com

The illustrations are torn- and cut-paper collage.

ISBN: 978-0-544-38760-7

Manufactured in China
SCP 10 9 8 7 6 5 4 3 2 1
4500742948

Houghton Mifflin Harcourt • Boston • New York

Frogs are creatures of two worlds—they spend part of their lives in the water and part on land.

These remarkable creatures have lived on earth for millions of years. In fact, a frog could have been stepped on by one of the first dinosaurs.

Today, frogs are found on every continent except Antarctica. Most live in or near the water, but some perch high in the treetops or hide on the forest floor. Others have found homes in caves, on desert sand dunes, or in people's houses. To survive in so many different habitats, frogs have evolved different ways of finding food, escaping danger, and attracting a mate. There are thousands of different kinds of frogs and they are found in an amazing variety of colors, shapes, and sizes.

An army of frogs

A group of frogs is called an **army** of frogs. So far, scientists have named more than 6,000 species of frog. And new kinds of frogs are being discovered all the time.

meowing night frog

Wallace's flying frog

tomato frog

lemur leaf frog

ornate horned frog

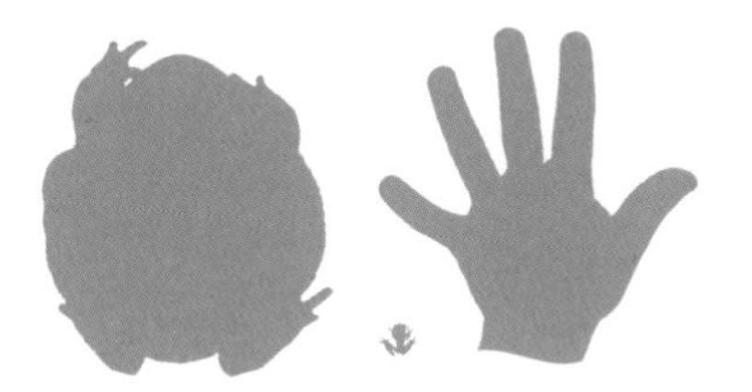

The frogs on this page are pictured at one-half life size. The **ornate horned frog** and the **meowing night frog** are shown above compared to an adult's hand.

Imbabura tree frog

long-nosed horned frog

Amazonian poison dart frog

crucifix frog

waxy monkey frog

What is a frog?

Frogs are amphibians, animals that can live both in the water and on land. Unlike mammals, which warm their bodies from within, frog's bodies are the same temperature as their surroundings. There are thousands of different kinds of frogs, but many of them share a few common traits.

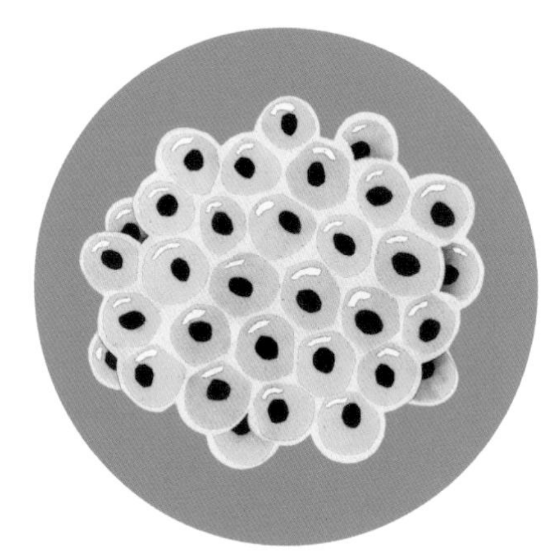

Most frogs lay their eggs—sometimes thousands of them—in water. The eggs stick together to form jellylike clusters.

Frogs don't usually drink water. Instead, they absorb it through their skin, which must be kept moist. Their skin also helps them absorb oxygen. The skin of some frogs contains powerful poisons.

Frogs that live in water have webbed toes to help them swim. Tree frogs use the sticky pads on the tips of their toes for climbing.

Long, strong back legs help frogs leap and swim.

A herpetologist (hur-pi-tol-uh-gist) is a scientist who studies amphibians and reptiles.

Many frogs have long, slim bodies.

The tympanum is a layer of skin that transfers sound and covers the frog's ears to keep out water and dirt.

When a frog swallows, its eyeballs sink into its head and help push food down its throat.

Many frogs have long, sticky tongues that can be flipped out to snag insects and other small animals.

Most frogs have teeth only on their upper jaw. Frogs don't chew their food—their teeth help them hold on to their prey until they swallow it.

Frog bodies are often darker on top and lighter on the bottom. When seen from above, the frog's dark back blends in with dark water or vegetation. When it's in the water and viewed from below, the frog's light belly makes it difficult to spot against a bright sky.

Some frogs use their hands to push their prey into their mouth.

Big bulging eyes on the top of its head allow a frog to see in almost every direction.

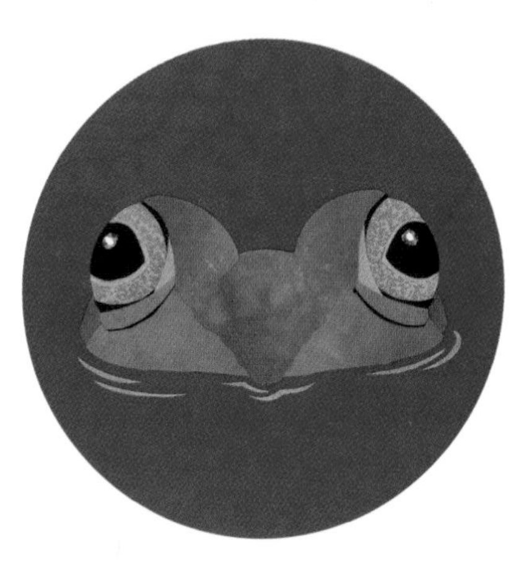

Frog or toad?

Frogs and toads are members of the same group of animals. To scientists, toads are simply one kind of frog. But there are some differences between toads and other frogs. Here are some of the features of a typical toad.

A group of toads is called a **knot** of toads.

Telling frogs and toads apart can be tricky. A few toads have smooth, slimy skin, and some frogs have dry, bumpy skin.

Toads have poison glands, or sacs, behind their eyes. These sacs ooze toxins into the mouth of any predator that bites down on the toad.

The toad's eyes are lower and more forward-facing than the frog's.

Toads don't have teeth.

The toad's tongue is shorter than a frog's, but it is still sticky and good for snagging insects. Most toads and frogs won't eat prey that isn't moving.

Most toads have dry, bumpy skin, so they don't have to live as close to water. Their skin also contains deadly toxins.

Toads have short stout bodies.

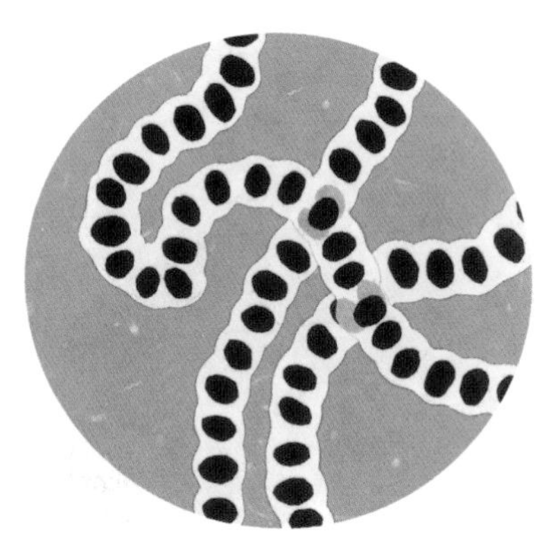

Toads lay their eggs in long strands, which stay in place by wrapping themselves around plants in the water.

The toad's toes aren't webbed. The toes on its back feet are often used for digging.

The toad's legs are shorter than a frog's. They are used to walk or hop, not to take long leaps.

Leading a double life

Most amphibians start life as larvae that live in the water and breathe with gills. As they grow, they go through metamorphosis, changing into four-legged adults that can live on land and breathe air. There are three main groups of amphibians.

Frogs and toads are by far the largest group of amphibians.

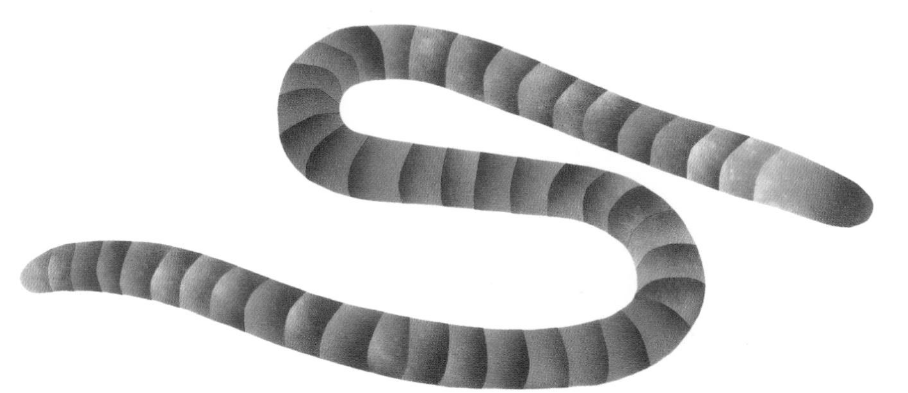

Caecilians don't have legs. They look like worms or snakes, but are rarely seen because they spend their lives underground. Caecilians live in tropical parts of the world. There are about 175 species of caecilian.

Salamanders look a bit like lizards. **Newts** are a kind of salamander—most of them have bumpy skin. They spend part of their lives on land and part in the water. There are more than 600 species of salamander.

Finding a mate

Frogs must find a mate to produce offspring. In the frog world, it's usually the male's job to attract a female. Some male frogs signal a female by changing color or using gestures, but most try to charm their potential mates with sound. They grunt, squeak, croak, and chirp. Each species of frog has its own special call.

The **smooth guardian frog** is the only frog we know of that reverses the usual mating ritual. It is the female of this species that calls out to the males.

The **Pinocchio frog** gets its name from its long nose. When a male calls out, his nose inflates and points upward.

For much of the year, the male **Indian bullfrog** is a dull green color. But when mating season arrives, he turns bright yellow. He forces air from his bright blue vocal sacs to create a loud croaking sound.

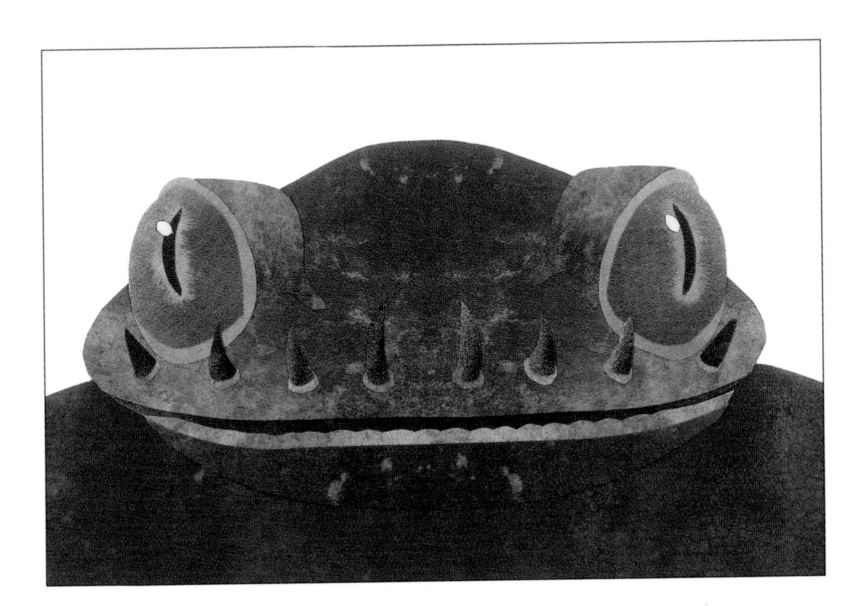

The male **mustache toad** uses the sharp spines on its upper lip to battle other males and win the rights to a nesting site—and the affection of a female. The spines fall off when the mating season ends.

The male **Surinam golden-eyed tree frog** uses his impressive vocal sacs to produce a loud mating call.

The call of the male **splendid leaf frog** is not very loud, so this frog signals potential mates by lifting one of his colorful legs and waving it back and forth.

It's all about the eggs

Many animals—including frogs—like to eat frog eggs. Some frogs lay thousands of eggs, then abandon them. Most will perish, but perhaps a few will survive. Other frogs protect and care for their eggs, often in surprising ways.

After a female **midwife toad** lays her string of eggs, her mate wraps them around his legs to keep them safe. When the eggs are ready to hatch, he wades into the water and lets the tadpoles swim away.

When a mother **Surinam toad** lays her eggs, the male toad presses them into the female's back. Her skin grows over the eggs, which develop into tadpoles. Several months later, dozens of little toads will emerge from the pits in their mother's back.

The male **three-lined poison dart frog** guards his mate's eggs until they hatch. Then he carries the tadpoles on his back to a stream or pond.

The male **Darwin's frog** protects his mate's eggs by carrying them in his special vocal sac. He keeps them safe until they have become little frogs. Then he opens his mouth, and as many as twenty froglets clamber out.

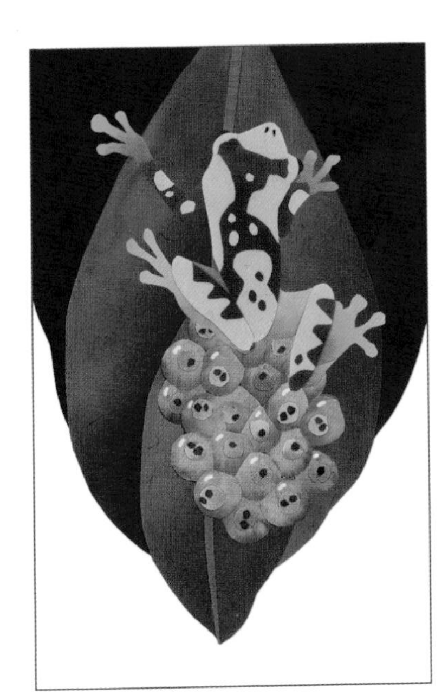

The **hourglass tree frog** lays her eggs on a leaf that overhangs a stream. As soon as the tadpoles emerge, they drop into the water.

The female **túngara frog** makes a nest of foamy bubbles. She lays her eggs inside and the male frog fertilizes them. The foam protects the eggs and keeps them moist until they hatch. After a while, the nest falls apart and the tadpoles drop into the water below.

Growing up

Like all amphibians, frogs change shape as they grow up. The transformation from egg to tadpole to frog is called **metamorphosis**.

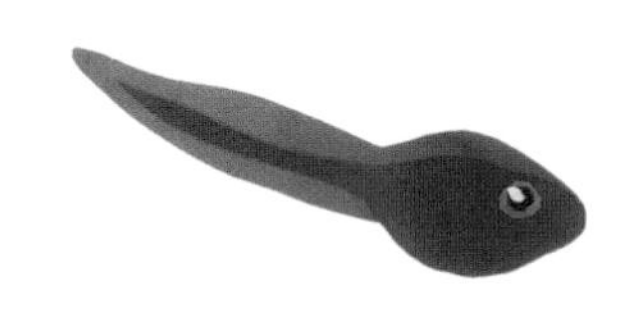

❸ The tadpole has a long tail. It can breathe underwater using internal gills. It feeds on algae and water plants.

❹ The tadpole grows, and bulges form where its legs will appear. Tadpoles are sometimes called "polliwogs."

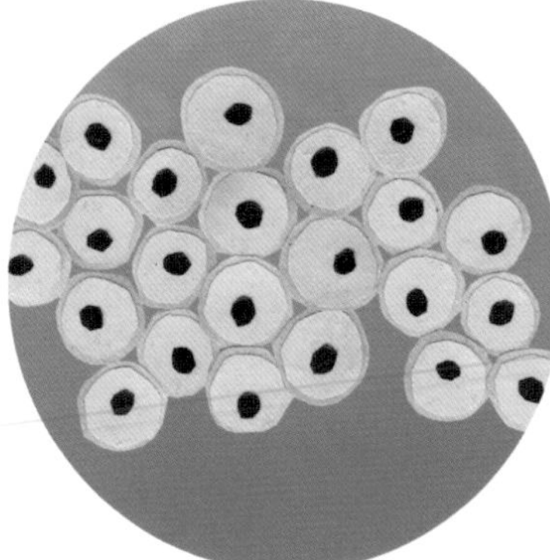

❶ A female frog usually lays her eggs in the water, where a male frog will fertilize them.

❷ Frog embryos feed off the yolk inside their eggs. After a few days or weeks, tadpoles begin to hatch from the eggs.

❺ A pair of back legs appear, and the tadpole begins to develop the lungs that it will need to breathe the air.

❻ Front legs emerge. The tadpole is starting to look more like a frog. It gets most of its nutrition by absorbing its tail.

The time from egg to adult frog varies quite a bit, depending on species. These figures are an average.

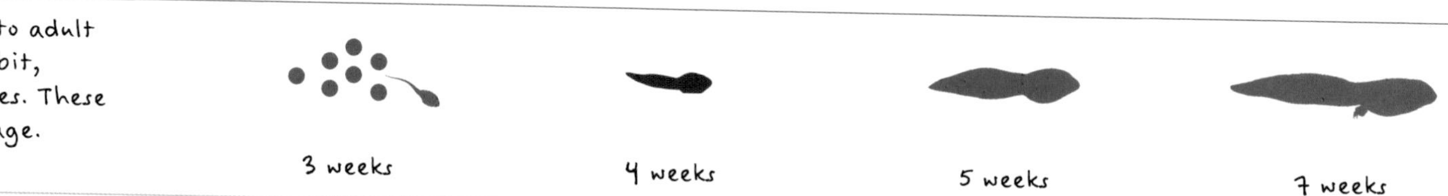

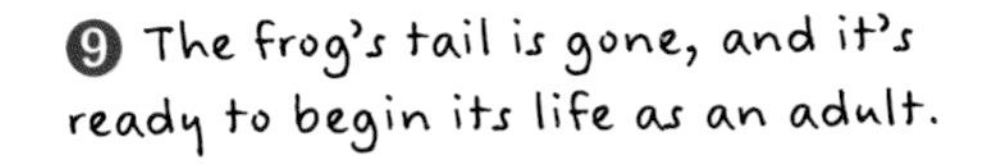

❾ The frog's tail is gone, and it's ready to begin its life as an adult.

❽ The froglet is almost an adult frog. It has absorbed most of its tail.

❼ The tadpole has become a froglet. It can breathe air, and its tail is getting smaller.

At home in the water . . .

Most frogs spend at least part of their lives in the water. But a few never leave. They are aquatic frogs, and they've adapted to a full-time—or nearly full-time—life underwater.

The **Titicaca water frog** (right) lives in a deep, cold lake in South America. It never leaves the water. Unlike most frogs, it does not always need to come to the surface to breathe—its folds of skin help it absorb oxygen from the water.

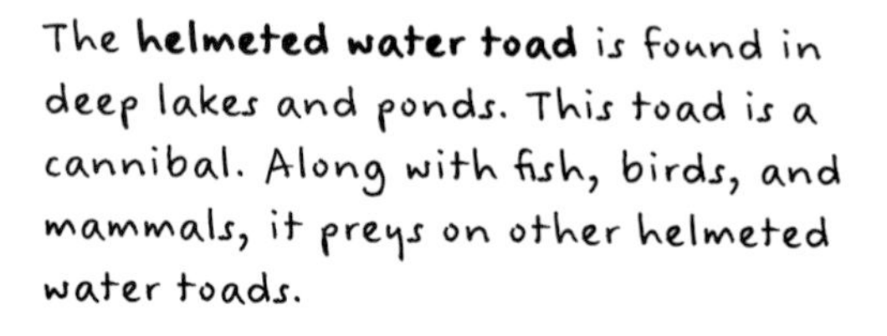

The **helmeted water toad** is found in deep lakes and ponds. This toad is a cannibal. Along with fish, birds, and mammals, it preys on other helmeted water toads.

The **African clawed frog** is a powerful swimmer. It doesn't have teeth or a tongue, so it uses its hands to push food into its mouth.

. . . and on land

Some frogs have found a home on the ground —or beneath it. Many of these frogs dig burrows, and a few stay deep underground for months or years, coming to the surface only after a hard rain.

The **water-holding frog** spends the dry months of the year buried deep in the sand. It wraps itself in a mucous cocoon, and it can store extra water in its body. Thirsty native people sometimes dig up these frogs and squeeze them gently to produce drinking water.

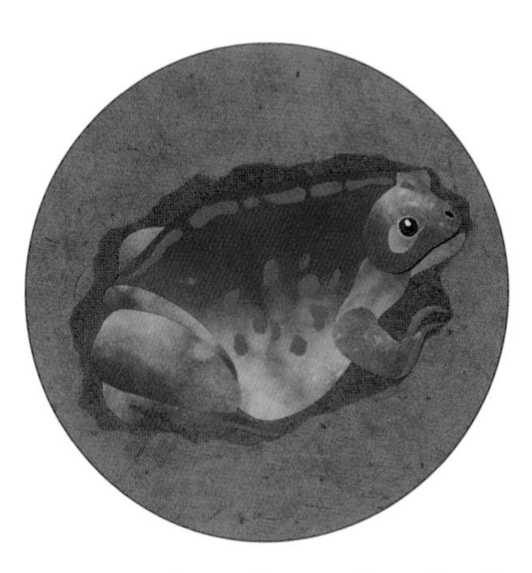

The **Mexican burrowing toad** digs itself backwards into the ground, using specially adapted back legs and feet. It spends most of its life underground, emerging to lay its eggs after it rains.

Using its nose as a shovel, the **spotted shovel-nosed frog** digs into mud and wet soil. It tunnels along beneath the surface as it searches for worms and insects to eat.

The **desert rain frog** lives on sand dunes near the ocean. It gets the water it needs from fog that drifts in off the sea. The call of this frog sounds like a squeaky toy.

Life in the trees . . .

One group of frogs has found that life is easier above the ground. Living in the treetops puts these frogs out of reach of many predators, but it requires some special adaptations.

Imbabura tree frogs are active at night. Their enormous eyes help them as they search for insects to eat in the dim light.

The **Amazon milk frog** lays its eggs in water-filled tree holes high above the ground. It defends itself with a poisonous white liquid that oozes from its skin.

The cricket glass frog sleeps during the day, using its sticky toe pads to attach itself to the underside of a leaf. It looks like a Muppet, a fact that has given it the nickname "Kermit frog."

. . . and elsewhere

These frogs make their homes in some unusual places.

The **banded bullfrog** lives in typical frog habitats such as forests and grasslands. But it's also found in houses in Southeast Asia. Most people don't mind sharing their homes with this frog, because it eats lots of pesky insects.

The **waterfall frog** lives near streams and often takes shelter behind waterfalls. It doesn't have vocal cords—they are not needed, because the noise of falling water would drown out any call.

The **wood frog** is the only amphibian that lives above the Arctic Circle. In winter, it nests in leaf litter or in an old log, then freezes solid. This frog has special chemicals in its body that allow it to withstand cold temperatures that would kill most amphibians.

The tiny **cave squeaker frog** lives in woods, fields, and caves. Its call is a high squeaky chirp.

What do frogs eat?

Frogs are carnivores—they eat other animals. Some frogs specialize and feed on one particular kind of insect. Others will gulp down anything that fits into their mouth, including other frogs. Most frogs will eat only live prey. Frog eyes are sensitive to movement. If their prey is motionless, frogs don't seem to recognize it as food.

The **purple frog** spends most of its life underground, where it feeds on termites. It searches for its prey as much as ten feet (3 meters) below the surface.

White's tree frog eats insects, spiders, small frogs, and mammals, including bats.

The **African bullfrog** is a voracious eater. Its diet includes insects, snakes, fish, birds, mammals, and frogs. It will even gulp down its own tadpoles.

The **crab-eating frog** is the only frog that can live, at least for short periods of time, in salt water. It eats insects, crabs, and other small animals.

The **giant monkey frog** eats with its hands instead of its tongue.

The **Amazon horned frog** appears to be mostly mouth. It is an ambush hunter, waiting for its prey to wander by, then lunging and swallowing it. People walking where this frog lives often wear boots—the frog sometimes leaps at hikers and tries to bite their ankles.

Frog defenses

Frogs have many enemies, but they have come up with some unusual—and effective—ways of defending themselves.

The **hairy frog** can break its own finger bones, which penetrate its skin and act as sharp claws. The filaments growing from the frog's body act as gills, allowing the frog to absorb more oxygen from the water.

The **Budgett's frog** is small, but it is not easily frightened. If it is attacked, this aggressive frog stands up tall and screams loudly. It will also lunge and bite.

Bruno's casque-headed frog is one of only two venomous* frogs (Greening's frog is the other). It injects toxins with spines on its upper lip. The powerful venom is said to be extremely painful, and it can be dangerous to a human.

*Note: Venomous animals inject their toxin with teeth, spines, or stingers. Poisonous animals have toxins in their skin or flesh, and must be touched or eaten to be dangerous.

The **pebble toad** clings to steep, rocky hillsides. It escapes danger by letting go and bouncing down the hillside like a pebble. It is so small and light that the fall does not hurt it.

This **leaf-litter frog** is "playing dead." Many predators prefer to eat live prey, so this can be an effective defense.

The **black rain frog** dwells in underground tunnels. If it is threatened, it inflates itself with air. This makes it difficult for a predator to pull the frog from its burrow.

Poison frogs

How do you protect yourself if you are small, soft-bodied, and not very fast? One good way is by being poisonous. A predator that tries to eat one of these frogs is making a serious mistake.

The tiny **mantella frog** lives among leaf litter on the forest floor. Its bright colors warn predators that it is toxic.

The **Mount Iberia frog** is one of the smallest frogs in the world. Its skin is also highly poisonous. Like most poison frogs, its toxins comes from the insects it feeds on.

 This is the actual size of the Mount Iberia frog.

There are two large poison-filled glands on the neck of the **cane toad**. Any predator that bites down on this toad gets a mouthful of deadly toxins. The poison works so quickly that snakes and wild dogs have been found dead with a cane toad sticking out of their mouth.

Few predators try to eat the **crucifix frog**—its skin is covered in an extremely sticky goo that would glue their mouth shut. Biting insects also become trapped in the frog's glue. About once a week it sheds its skin and eats it, bugs and all.

If it is attacked, the bright red **tomato frog** swells up and secretes a sticky, poisonous mucous. This toxic slime burns the mouth and eyes of any animal that tries to eat this frog.

Startle, hide, and mimic

Some frogs defend themselves with poison. Others swell up, ooze sticky mucus, or fight back. But a few frogs have come up with trickier ways of protecting themselves. They use camouflage, mimicry, or the element of surprise to escape danger.

It sleeps with its eyes closed and its legs tucked beneath it. But if danger threatens, the **red-eyed tree frog** pops open its enormous eyes and displays its colorful feet. This sudden flash of color can startle a predator long enough for the frog to get away.

The **gray tree frog** is a master of camouflage, changing color to match its surroundings. It can turn itself from green to gray to brown to almost white.

As it perches on a rock or tree trunk, the **Vietnamese mossy frog** appears to be a clump of moss.

The **pied warty frog** is also known as the bird poop frog. Predators are likely to mistake it for bird droppings on a leaf.

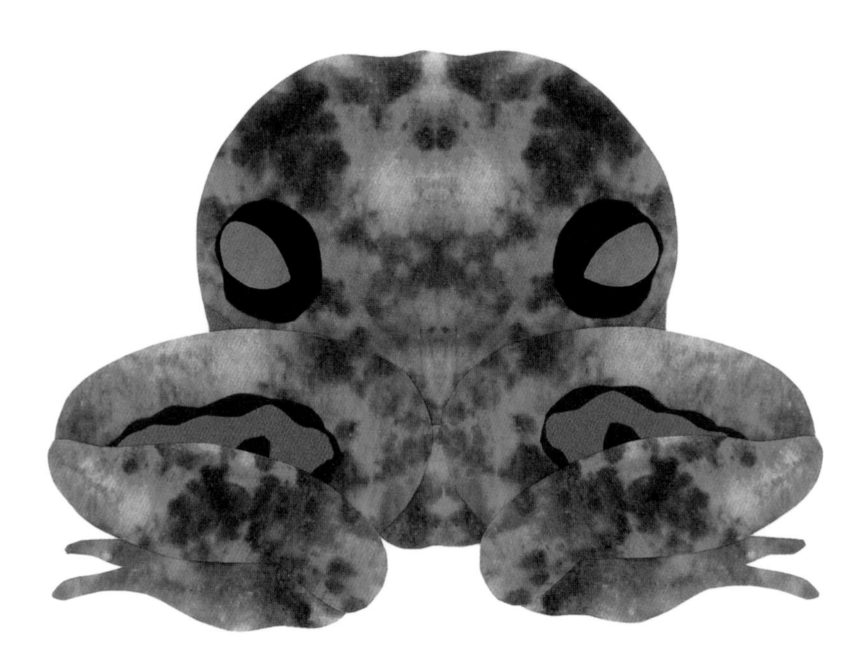

What does the **Colombian four-eyed frog** do if a bird or snake threatens it? It turns around and lifts its rear end, revealing two poison glands that look like big eyes. A predator may decide to look for a meal somewhere else. If it does attack, it may find itself with a mouth full of poison.

Despite its name, the **Zaparo's poison frog** (above) is not poisonous. It mimics a colorful poison dart frog, fooling its enemies.

The **long-nosed horned frog** (left) looks almost exactly like a dead leaf. As long as it doesn't move, it's almost impossible to spot as it sits on the forest floor.

Extreme frogs

Here are a few of the record holders in the frog world.

The **golden poison frog** is a kind of poison dart frog. It is small, but its skin contains the most powerful poison of any frog. One frog contains enough poison to kill ten full-grown humans.

The **paradoxical frog** is also known as a shrinking frog. It begins life as a huge tadpole. The adult frog is much smaller—the frog and tadpole above are shown at the same relative size.

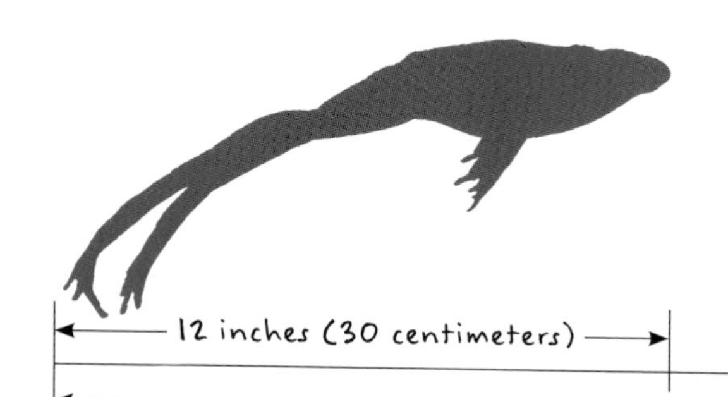

The **cane toad** can lay as many as 30,000 eggs at a time. But the **Noble's pygmy frog**, named for its small size, lays as few as two eggs at a time.

The **common coqui frog** may have the loudest call of any frog. Its call is louder than a lawnmower.

In the wild, most toads probably live no more than 10 or 12 years. But one **common toad** in Great Britain has lived in a garden for almost 40 years.

The **American bullfrog** can jump more than seven feet in one leap.

A recently discovered **New Guinea frog*** holds the record for both smallest frog and smallest animal with a backbone. The **goliath frog** is the world's largest. Both frogs are shown life-size.

* So far this frog has only a scientific name: *Paedophryne amauensis.*

7 feet (213 centimeters)

Frogs in danger

One-third of all frog species are in danger of extinction. They are threatened by a warming climate, pollution, the destruction of their habitat, and a deadly fungus that has spread through many frog populations. Frogs and other amphibians absorb air and water through their skin, so they are easily harmed by pollutants in the air or water. Sick or dying frogs are often the first warning of environmental problems.

The frog is often described as a "canary in a coal mine." Canaries are especially sensitive to dangerous gases in the air they breathe. Coal miners used to take a canary with them into the mines. If the canary died, the miners knew they should get out right away. Today the term "canary in a coal mine" is used to describe something that gives an early warning of danger.

Despite its name, the colorful **clown frog** is actually a toad. It lives in the rainforests of Central America, and it is critically endangered. Frogs are helpful predators, consuming many insects that we consider pests. They are also an important food source for many larger animals, including snakes, birds, mammals, and fish.

Gone forever

These frogs are extinct—they are never coming back.

The **golden toad** lived in one small part of the Costa Rican rainforest. The last golden toad was seen in 1989.

The mother **gastric brooding frog** incubated her eggs in her stomach, something no other frog does. These frogs lived in Australia, and were last seen in the 1980s.

Rabb's fringe-limbed tree frog was a kind of flying frog. It used the webbing on its large hands and feet to glide from tree to tree. This frog was last seen in the wild in 2007. The last of these frogs lived in a zoo. It was named Toughie, and it died in 2016.

A success story?

The **California red-legged frog** is still in danger. But some of its habitat is now protected and its eggs have been introduced to new environments, giving it a chance to make a comeback.

This table shows the size, diet, and range of the frogs in this book. Body length is measured from snout to vent, which means from the tip of the nose to the end of the body, not including the legs. With just a few exceptions, female frogs are larger than males of the same species. Average female lengths are given here.

	name	body length	diet	range
cover	**green poison dart frog**	2 in. (5 cm)	ants and beetles	southern Central America, NW South America
page 4	**meowing night frog**	½ in. (1.2 cm)	mites, small insects, spiders	India
page 4	**Wallace's flying frog**	3½ in. (9 cm)	insects	Indonesia
pages 4, 25	**tomato frog**	3 in. (7.5 cm)	insects, frogs, rodents	Madagascar
page 4	**lemur leaf frog**	1½ in. (3.8 cm)	insects, worms, slugs	Central America, northern South America
page 4	**ornate horned frog**	6½ in. (16.5 cm)	insects, spiders, small reptiles, mice	southern South America
pages 5, 18	**Imbabura tree frog**	2½ in. (6.4 cm)	insects, spiders	northwest South America
pages 5, 27	**long-nosed horned frog**	4½ in. (11.4 cm)	insects, spiders, mice, lizards, frogs	southern South America
page 5	**Amazonian poison dart frog**	¾ in. (2 cm)	ants, beetles	Central America, northern South America
pages 5, 25	**crucifix frog**	2½ in. (6.4 cm)	ants and termites, other insects	eastern Australia
page 5	**waxy monkey frog**	3 in. (7.5 cm)	insects	South America
page 10	**Indian bullfrog**	1½ in. (3.8 cm)	insects, worms, frogs, rodents, birds	India, Bangladesh, northern Pakistan
page 10	**smooth guardian frog**	1½ in. (3.8 cm)	insects, small invertebrates	Indonesia
page 10	**Pinocchio frog**	1¼ in. (3.2 cm)	insects	Indonesia
page 11	**mustache toad**	3 in. (7.5 cm)	insects, frogs, rodents	China
page 11	**Surinam golden-eyed tree frog**	2½ in. (6.4 cm)	worms, insects, small fish	South America
page 11	**splendid leaf frog**	3¼ in. (8.3 cm)	insects, spiders	Central America, northern South America
page 12	**midwife toad**	2¼ in. (5.7 cm)	insects, spiders, worms	Europe, northwestern Africa
page 12	**Surinam toad**	2½ in. (6.4 cm)	worms, insects, small fish	South America
page 12	**three-lined poison dart frog**	2¼ in. (5.7 cm)	insects	northern South America
page 13	**Darwin's frog**	1 in. (2.5 cm)	insects, worms, snails, spiders	Chile, Argentina
page 13	**hourglass tree frog**	1½ in. (3.8 cm)	moths, other insects	southern Mexico, Central America
page 13	**túngara frog**	1¼ in. (3.2 cm)	ants, termites, worms	Central America, northern South America
page 16	**Titicaca water frog**	5¼ in. (13.3 cm)	snails, insects, tadpoles, fish	Peru, Bolivia
page 16	**helmeted water toad**	12½ in. (32 cm)	insects, fish, frogs, birds, mammals	Chile
page 16	**African clawed frog**	3 in. (7.5 cm)	insects, fish, worms, carrion	central Africa
page 17	**desert rain frog**	1¼ in. (3.2 cm)	beetles, insect larvae	Namibia, South Africa
page 17	**Mexican burrowing toad**	3 in. (7.5 cm)	ants, termites, other insects	southern Texas and Central America
page 17	**water-holding frog**	2¾ in. (7 cm)	insects, small fish	southern Australia
page 17	**spotted shovel-nosed frog**	3 in. (7.5 cm)	termites, earthworms	South Africa
page 18	**Amazon milk frog**	4 in. (10 cm)	worms, crickets, other insects	northern South America
page 18	**cricket glass frog**	1 in. (2.5 cm)	insects	Central America, South America
page 19	**banded bullfrog**	3 in. (7.5 cm)	flies, crickets, other insects, worms	Southeast Asia
page 19	**waterfall frog**	2 in. (5 cm)	insects, worms, aquatic invertebrates	northern Australia
page 19	**wood frog**	2¾ in. (7 cm)	insects, spiders, worms	Canada, northeastern United States
page 19	**cave squeaker frog**	1 in. (2.5 cm)	insects, spiders, worms	Zimbabwe
page 20	**purple frog**	2¾ in. (7 cm)	termites	India
page 20	**White's tree frog**	4 in. (10 cm)	insects, spiders, small frogs, bats	Australia and southern New Guinea
page 20	**African bullfrog**	9 in. (23 cm)	insects, rodents, reptiles, birds	central and southern Africa
page 21	**crab-eating frog**	3 in. (7.5 cm)	insects, crabs, other crustaceans	Southeast Asia
page 21	**giant monkey frog**	4½ in. (11.4 cm)	insects	northern South America
page 21	**Amazon horned frog**	8 in. (20 cm)	frogs, insects, rodents, birds	northern South America

	name	body length	diet	range
page 22	**hairy frog**	4¼ in. (10.8 cm)	slugs, millipedes, spiders, insects	central Africa
page 22	**Budgett's frog**	4 in. (10 cm)	frogs, insects, snails	South America
page 22	**Bruno's casque-headed frog**	3¼ in. (8.3 cm)	insects, spiders, centipedes	southeastern Brazil
page 23	**leaf-litter frog**	2 in. (5 cm)	spiders, grasshoppers, other insects	southern Brazil and northern Argentina
page 23	**black rain frog**	2 in. (5 cm)	insects, spiders, millipedes	South Africa
page 23	**pebble toad**	1¼ in. (3.2 cm)	insects	Venezuela
page 24	**mantella frog**	1¼ in. (3.2 cm)	ants, beetles, mites	Madagascar
page 24	**Mount Iberia frog**	½ in. (1 cm)	insects, insect larvae, mites, spiders	eastern Cuba
pages 24, 28	**cane toad**	9 in. (15 cm)	insects, frogs, rodents	Central and South America, Australia
page 26	**red-eyed tree frog**	2½ in. (6.4 cm)	crickets, moths, flies, other insects	Central America
page 26	**gray tree frog**	2 in. (5 cm)	insects, spiders, mites	eastern United States and Canada
page 27	**Vietnamese mossy frog**	3½ in. (9 cm)	crickets, cockroaches, earthworms	northern Vietnam
page 27	**pied warty frog**	1¼ in. (3.2 cm)	insects	India, Southeast Asia, Indonesia
page 27	**Colombian four-eyed frog**	1¾ in. (4.5 cm)	insects, spiders	Panama, northern South America
page 27	**Zaparo's poison frog**	1½ in. (3.8 cm)	ants, beetles, small invertebrates	Ecuador and Peru
page 28	**golden poison frog**	2 in. (5 cm)	ants, crickets, beetles, termites	Colombia
page 28	**paradoxical frog tadpole**	10 in. (25 cm)	algae, water plants	South America
page 28	**paradoxical frog**	2½ in. (6.4 cm)	larvae, insects, aquatic invertebrates	South America
page 28	**Noble's pygmy frog**	½ in. (1.2 cm)	insects, spiders, mites	South America
page 28	**common coqui frog**	1¾ in. (4.5 cm)	spiders, moths, crickets, snails, frogs	Puerto Rico, Hawaii
page 28	**common toad**	6 in. (15 cm)	insects, spiders, earthworms, mice	Europe, North Asia, northwest Africa
page 28	**American bullfrog**	6 in. (15 cm)	insects, spiders, fish, reptiles, mice	North America, South America, Asia
page 29	**New Guinea frog**	⅜ in. (1 cm)	small insects, spiders, mites	New Guinea
page 29	**goliath frog**	12½ in. (32 cm)	insects, spiders, worms, reptiles, bats	Central Africa
page 30	**clown frog**	1¾ in. (4.5 cm)	ants, other insects, spiders	Costa Rica, Panama
page 31	**golden toad**	2 in. (5 cm)	insects, spiders, worms, millipedes	Costa Rica, Central America
page 31	**Rabb's fringe-limbed tree frog**	3½ in. (9 cm)	crickets, cockroaches, other insects	Panama
page 31	**gastric brooding frog**	3 in. (7.5 cm)	crayfish, frogs, insects	eastern Australia
page 31	**California red-legged frog**	5 in. (12.7 cm)	insects, fish, frogs, mice	California, northwestern Mexico

For more information:

Books

Everything You Need to Know About Frogs and Other Slippery Creatures. Edited by Carrie Love. DK Publishing, 2011.

Face to Face with Frogs. By Mark Moffett. National Geographic, 2008.

Frog. By Charlotte Sleigh. Reaktion Books, 2012.

Frogs. By Seymour Simon. Harper Collins, 2015.

Frogs, Toads, and Turtles. By Diane L. Burns. Northwood Press, 1997.

The Frog Scientist. By Pamela S. Turner. Sandpiper, 2009.

The Mystery of Darwin's Frog. By Marty Crump. Boyd's Mill Press, 2013.

Red-Eyed Tree Frog. By Joy Cowley. Scholastic Press, 1999.

Watch Me Grow: Frog. By Lisa Magloff. DK Publishing, 2003.

Websites

Amphibiaweb
amphibiaweb.org

AnimalSake
animalsake.com/types-of-frogs

Defenders of Wildlife
defenders.org/frogs/basic-facts

LiveScience
livescience.com/50692-frog-facts.html

National Geographic
video.nationalgeographic.com/video/animals/amphibian

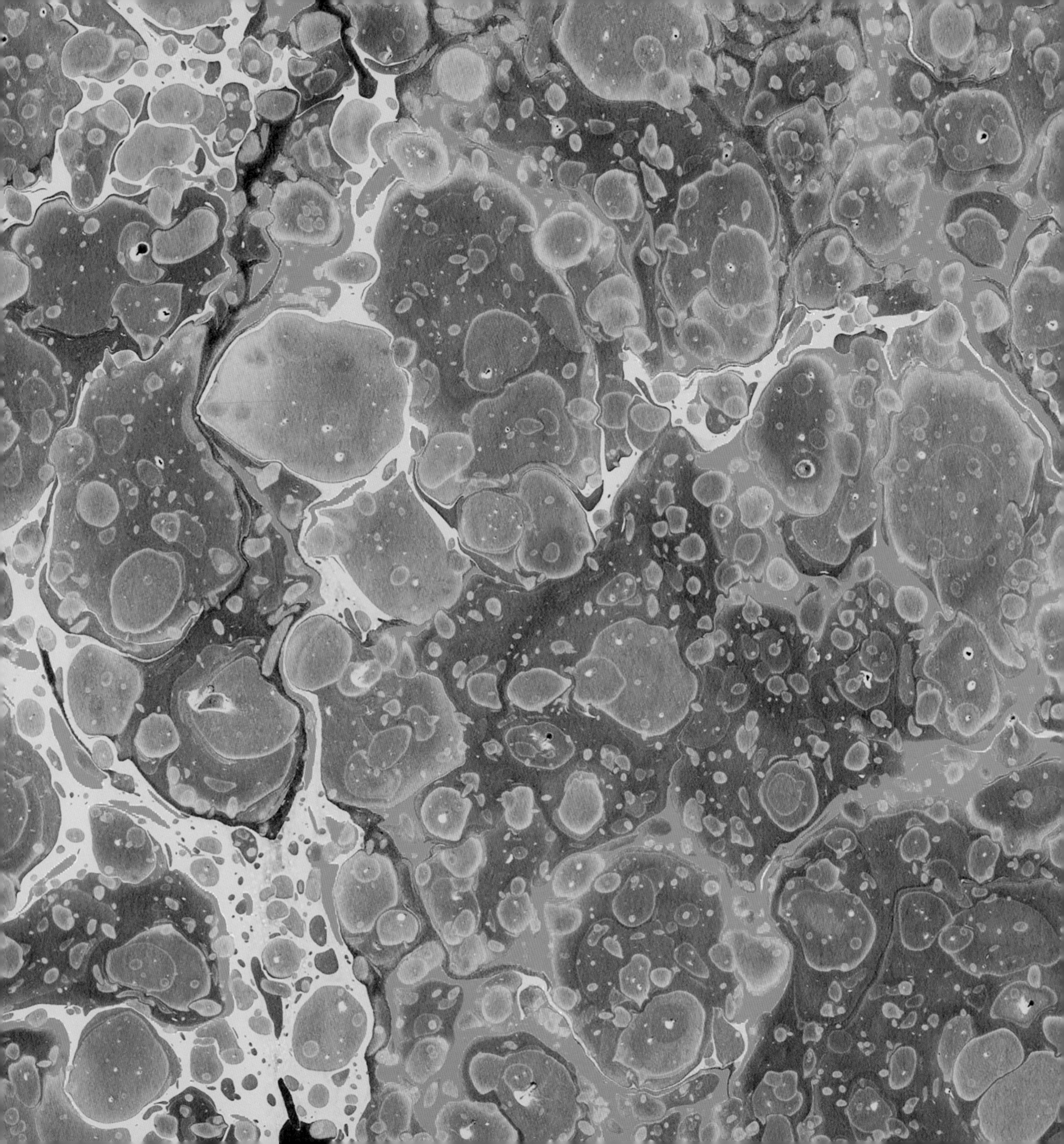